LIVERWORT TERMINOLOGY

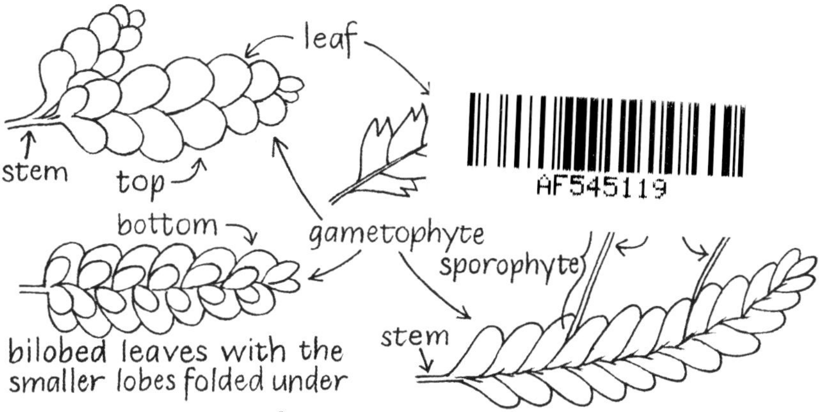

Leafy Liverworts

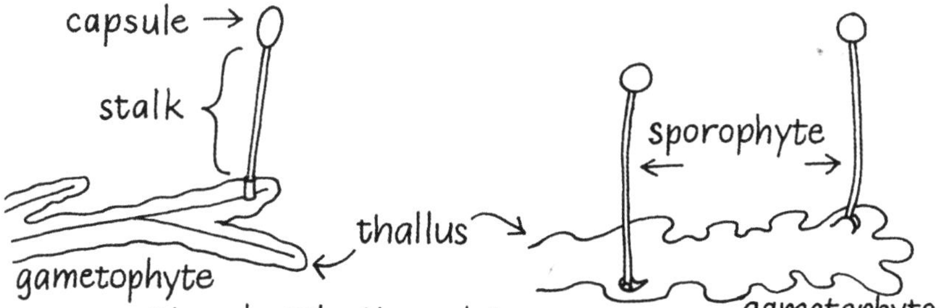

Simple Thallose Liverworts

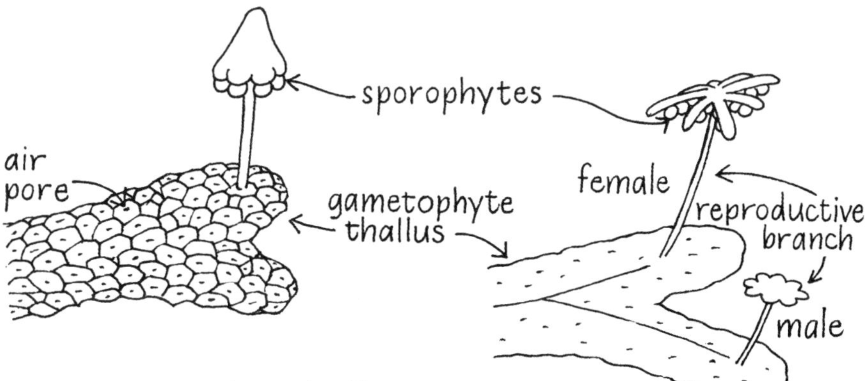

Complex Thallose Liverworts

A liverwort and several kinds of mosses on a streambank

Outstanding

MOSSES & LIVERWORTS

of Pennsylvania
& Nearby States

Susan Munch

Albright College

This book is dedicated to all the people who have taught me about Bryophytes: especially L. G. Livingston (1905-1988) and Elva Lawton (1896-1993), and to all the students, friends, and relatives who have encouraged me by their interest in mosses and liverworts.

To order this book, e-mail the author:
susanm@alb.edu

Printed with the assistance of a grant from the Pennsylvania Wild Resources Conservation Program.

My best thanks to Rob Fergesen, Dick Heller, Naja Kraus, Carolyn Reider, and Jack Stabley for help with the photographs; to Sharon Began, Larry Klotz, John Kunsman, Carol Loeffler, and Christine Manville for botanical advice; to Jon Bekken, Marissa Moyer, and Heather Urkusky for computer work; to Nancy Kerper for secretarial help; and especially to my brother Charles Munch for book design, line-drawings, and proof-reading.

All photographs are by the author except where the name of the photographer is indicated. The magnifications with the photographs are approximate.

Cover – Plagiochila asplenioides, a leafy liverwort

Outstanding Mosses and Liverworts of Pennsylvania & Nearby States

Table of Contents

Neckera pennata is a rare moss found in Pennsylvania on hardwood trees among hemlocks. The transverse waves in the leaves and the nearly horizontal position of the branches extending out from the tree's bark make this species especially distinctive. x5

PREFACE

This guide is intended for nature-lovers in Pennsylvania and neighboring states who want an introduction to the local mosses and liverworts but do not have the time or inclination to learn all the technical details used by botanists to identify the species. It makes the recognition of some species or genera of mosses and liverworts possible without the use of a microscope. Included in the guide are about fifty kinds of mosses and liverworts that are very common, very distinctive, or very large.

However, even large mosses are fairly small plants. Some people with really good close eyesight can see enough detail by holding the plants close to their eyes, but most need an 8x or 10x hand-lens (or loupe or magnifying glass) to examine the distinctive features of mosses and liverworts.

There are over three hundred species of mosses in Pennsylvania and over one hundred of liverworts. Some of these stand out distinctly from the others in our forests and wetlands, and a few are especially noticeable in cities. The mosses and liverworts that frequently catch the enquiring eye are illustrated and described in this book. In some cases individual species are easily recognized, while in others several similar species must be grouped together and recognized only at the genus level. Most major genera are covered, the exceptions usually being cases where recognition is nearly impossible without much greater magnification. Readers who have identified many of the species covered in this guide and have begun noticing others not included can move on to more technical sources listed on p. 88 if they have access to a compound microscope (i.e., magnification of 400x).

Although the technical details and volume of terminology will be kept to a minimum in this book, some understanding of the structure and reproduction of mosses and liverworts is necessary to identify them. The users of this guide are advised to read the following introductory sections on life-cycles, structure, and growth, before going on to the descriptions of specific mosses and liverworts. A glossary (p. 86-87) and labeled drawings of mosses and liverworts (inside the front and back covers) are also included to help those who are unfamiliar with moss and liverwort vocabulary.

Moss with mature sporophytes x2

Leafy liverwort with mature sporophytes x4

INTRODUCTION TO MOSSES AND LIVERWORTS

Life Cycles of Mosses and Liverworts

Mosses and liverworts are small delicate land plants usually considered to be related. Together they are referred to as bryophytes. Mosses and liverworts are classified in the same larger group because they share a very unusual life cycle. All bryophytes have two very different forms during two stages of their life. The gametophyte stage is the green moss or liverwort you see every day. The sporophyte stage, which is attached to certain mature gametophytes, is rarely green at maturity and normally stands above the green gametophytes. The sporophyte stage is temporary, while the gametophyte grows for months before the sporophytes form and normally continues growing after they are gone. When a gametophyte has grown long enough to be mature, sporophytes are produced as a result of sexual reproduction. Under the right conditions, the gametophytes produce eggs and sperm in separate microscopic reproductive organs. The sperm swim across the wet surface of the plant to the eggs, where fertilization occurs within the organ that produced the egg.

Each fertilized egg (or zygote) is retained in the reproductive organ that produced it. This zygote grows into a sporophyte while still attached to the green gametophyte. The sporophytes are normally brown or colorless at maturity and are shorter-lived than the gametophytes. The spores released by the sporophytes grow into new gametophytes if they land in a suitable location. In some cases, asexual reproduction may also take place, allowing the bryophytes to form new gametophytes without going through the sporophyte stage. Small green buds (or gemmae) may be produced on branches or in splash-cups. These gemmae are knocked loose from the parent plant and grow into independent individuals. Fragments of the gametophyte plant may also grow into new plants.

Structure and Growth

All bryophytes have a green gametophyte stage. This photosynthetic plant is fastened to its substrate by thread-like rhizoids. It has no roots to take up water and little or no ability to transport

Naja Kraus

Moss gametophytes with spirally arranged leaves x1

Moss with immature sporophytes x2.5

Moss sporophytes with calyptras x4

With calyptra and with operculum x4

With operculum at the black arrow x4

water upward. The plants generally take in carbon dioxide, water, and mineral nutrients over their whole surface as long as the air is moist. Under these conditions, they continue growing at the tips of their stems and stem branches. When the air is not moist, (since they have no waxy coating like higher plants and no roots) the plants dry out, stop growing, and often curl up somewhat. Bryophyte sporophytes have the function of releasing spores into the air to be blown away and start new gametophyte plants in new locations. For this reason the sporophytes are normally elevated above the gametophytes; they usually have an elongate stalk holding up the spore-producing capsule.

Moss gametophytes have tiny leaves that are usually spirally arranged or more rarely pushed to either side of the stem. The stems may be short and upright, or they may creep horizontally. Most commonly the leaves are long, narrow, and pointed, with a thickened supporting structure called a midrib running along the center of the leaf from its attachment to the stem towards the point.

The sporophytes of mosses elongate gradually before the spores are made and often grow tall, green, and very narrow, until the capsule swells near the top in preparation for spore production. The top of the immature sporophyte is usually covered by a delicate scrap of tissue called the calyptra (torn from the reproductive organ that enclosed the zygote). After the spores mature and the calyptra dries up and blows away, the capsule of the sporophyte is completely visible, topped by a lid or operculum that falls off to allow for spore release. Under the operculum most mosses have specialized structures called peristome teeth to regulate the release of the spores. Except in a few unusual genera, the peristome teeth move with changes in humidity, spreading when dry (to allow more spore release when the air is in a state to distribute the spores more effectively) and closing when wet. After spore release, the empty brown sporophytes may hang on for months.

Liverwort gametophytes may have leaves or they may be thallose; that is, they may consist of just a sheet of green tissue not differentiated into leaves and stems. It is easy to recognize a thallose liverwort, but most people have trouble distinguishing leafy liverworts from creeping mosses. Leafy liverworts have leaves that are either rounded or lobed. These are found on either side of the

A liverwort showing overlapping leaves in two rows x3

Four types of liverwort leaves and a moss leaf

Sphagnum x1

Carolyn Reider

Tetraphis showing splash-cups with gemmae x2

stem, and each leaf normally overlaps the adjacent one so that they are arranged like shingles on a roof. Liverwort leaves do not have midribs, and, if they have points, there are two or more per leaf, not one as in mosses. Thus with close examination it is not too hard to tell which creeping leafy bryophytes are liverworts and which are mosses. Liverwort sporophytes are very rarely seen. The immature sporophytes are usually hidden by the gametophytes for the time it takes the spores to develop. Then the stalk elongates rapidly, the capsule splits open (usually into four segments), and the spores are released, all in a few days at most. (See p. 2 for a picture showing the black spore-filled capsules and the empty brown split ones.) Normally, after spore release, the delicate white stalks quickly collapse and the capsules disintegrate.

Groups of Mosses

This book gives examples from five different groups of mosses. The first two include only one major genus each, and the third has two genera. The fourth is a little larger with three major genera, and the fifth is very large and diverse. Features of the sporophytes, especially the peristome teeth, are very important in defining these groups of related mosses. For convenience in organizing the descriptions of the different kinds of mosses in this guide, the first four groups, which have very unusual features, will be considered in one section, and the large group of ordinary mosses will be divided into two sections based on their pattern of growth.

1) Mosses in the genus <u>Sphagnum</u> all have leaves with water storage cells held in a network of photosynthetic cells. They are rather large mosses with upright stems and many branches crowded together near the growing tip of the plant. After sexual reproduction the sporophytes are elevated on special branches of the gametophyte, rather than on their own stalks. They have no peristome teeth. As it dries, the capsule shrinks, building up pressure inside that pushes the operculum off and shoots out the spores.

2) <u>Tetraphis</u> is an unusual small moss topped by sporophytes with four large, rigid peristome teeth. The gametophytes also often reproduce asexually by means of gemmae released by splash-cups formed by the uppermost leaves of the upright stems. The gemmae are small buds which grow into new gametophytes when raindrops knock them to the ground.

Diphyscium showing the white-tipped capsules x3

Two different species of Polytrichum x0.5

Jack Stabley

Ordinary moss capsules with 16 teeth x7

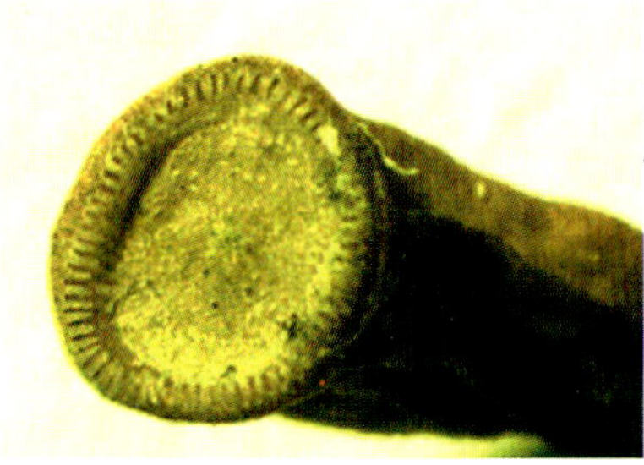

Jack Stabley

Upper end of a capsule of Polytrichum showing 64 teeth x15

A large complex thallose liverwort x2

Carolyn Reider

A simple thallose liverwort x2

3) Diphyscium and Buxbaumia have large asymmetrical capsules with reduced peristome teeth. A white, pleated cone with a small hole at the tip regulates the release of their spores.

4) Polytrichum, Atrichum and Pogonatum are related upright mosses with long pointed leaves having extra photosynthetic tissue attached to their upper surface. This extra tissue is like a series of ribbons running on edge along the surface of the leaf parallel to the midrib. When the operculum falls from the sporophyte, 32 to 64 very small, stiff peristome teeth are revealed. Since the capsule has only a circular slit that opens around its upper edge and the peristome teeth cover the slit at regular intervals, the spores sift out gradually as from a pepper shaker.

5) Most ordinary mosses have 8 or 16 peristome teeth that move in response to changes in humidity. Their leaves normally have only a single flat layer of photosynthetic cells. Some of these mosses grow upright and are unbranched or very little branched. (See pp. 28-44.) Their reproductive organs and sporophytes are at the tips of their stems. The plants often grow packed together in clumps. When a particular plant has produced a sporophyte, it often dies and is replaced by a new sister plant growing from the base of the clump. In contrast to these upright (or acrocarpous) plants, other mosses in this group are creeping and repeatedly branched (or pleurocarpous). Since pleurocarpous mosses grow with many sets of branches reaching out from the original stem, they often form large patches. These creeping mosses (see pp. 46-64) have their reproductive organs and sporophytes on side branches. When a particular side branch has produced a sporophyte, it may die and be overtopped by a younger branch.

Groups of liverworts

Of the two main forms of liverworts (leafy and thallose), the leafy liverwort group is the most diverse, with about a hundred species in Pennsylvania. Most are very small and grow on rotting logs or other very moist substrates. There are two groups of thallose liverworts, simple and complex. A few species of delicate simple thallose liverworts are often encountered in damp areas. They have little differentiation of tissues. The larger thallose liverworts are much more complex with several tissue types and a system of air-pores to admit carbon dioxide into the photosynthetic areas.

A moss that is restricted in Pennsylvania to dry calcareous rock – Rhytidium rugosum on a cliff-top x1.5

A living but dormant moss high on a dry tree-trunk x2

The exceptional leafy liverwort Frullania in its dark winter coloration, high on a tree trunk x3

The commonest of these live on rocks near streams. There are about thirty species of thallose liverworts in Pennsylvania.

Ecology

Because of their delicate structure and lack of a waxy cuticle, most bryophytes live in moist places or in drier areas in very sheltered microhabitats at ground level. Most of the bryophytes in this book grow in damp forests near streams. Some of the mosses remain dry and dormant for weeks or even months without much damage, and thus can live in dry forests or even out in the open. Since very few liverworts can tolerate desiccation, they are rarely found in dry locations. The leafy liverwort genus Frullania, which grows high up on trees, is the only real exception to this rule.

Each moss or liverwort species has specific substrates where it grows. Many are found on rocks, others on rotting logs, and some on bare soil. Some are restricted to one substrate, such as tree bark. The acidity (pH) of the rock or soil is often important. Some mosses are restricted to calcareous rocks like limestone and others to more acidic rocks like sandstone. Bryophytes that live on bare surfaces where larger plants can't grow act as pioneers in the plant community, gradually building up organic material for soil formation.

Though most bryophytes live in damp places on land, a few have become truly aquatic. For example, one of the mosses in this book (Fontinalis) is found in fast-flowing rocky streams. Many live in wetlands, the most important being members of the genus Sphagnum, which form peat bogs. In these bogs the level of acidity is high enough to slow the growth of fungi and bacteria responsible for decomposition. Lack of oxygen in peat further slows decay so that plant nutrients are less available. This limits the growth of many plants that might compete with the Sphagnum.

In contrast to these aquatic and wetland species, some mosses live on very dry rocks in wild habitats or in the city on rock walls or between paving blocks. A couple of these mosses associated with dry rocks or concrete are included in this book. Other mosses that can tolerate a lot of drying live high on the trunks of trees or even on their branches. These epiphytic mosses are usually very small and not often noticed.

A true moss with green leaves beside a greenish-gray lichen, sometimes called "reindeer moss" x1.5

A true moss, Mnium cuspidatum, growing with a lesser club-moss (Selaginella apoda) in a moist shady lawn x2

FINDING MOSSES & LIVERWORTS TO IDENTIFY

In Pennsylvania, the largest diversity of mosses and liverworts is found in damp forests along streams or near bodies of water. In these areas where the bryophytes are thriving, identification is easier than where they are too dry or otherwise stressed. For example, a large, moist, growing clump of a moss with sporophytes is easier to recognize than a desiccated bit of the same species in a location too dry to allow it to grow well or to produce sporophytes. Some mosses in this book can be easily identified without sporophytes, but, for most, features of the sporophyte help confirm the identity. (Liverworts, on the other hand, are identified almost completely with gametophyte features.)

The cooler seasons of the year are usually the best for looking at bryophytes – autumn through early spring, when rain is plentiful and the weather not too hot. Even when it is very cold, bryophytes by spring-fed streams are often in good condition. Don't try to use this book to identify other plants similar to bryophytes. It covers only true mosses, not "reindeer moss", which is a lichen, or any of the club-mosses, which are higher plants. Lichens can be distinguished from mosses by their gray-green outer layers covering white or black inner layers. The leaves of club-mosses are an opaque grass-green, rather than the translucent darker green of a typical bryophyte leaf. (See lower photo.)

Remember – there are many more mosses and liverworts in Pennsylvania than in this book. The descriptions and pictures that follow are intended to allow you to name the most easily identified bryophytes. When you have located as many mosses and liverworts as you can with this guide and want to learn more, try a more advanced book, one of those listed on p. 88, or a newer guide which will surely be available soon.

Meanwhile, if you have read the preceding introductory section, you are ready to start learning some mosses and liverworts. Begin by paging through the photographs in the following sections and then taking the guide (and a lens, if necessary) into the forest to look for the species shown. When you think you have found a match for a photograph, read the text to be sure that what you see fits the description. If you want a sample to examine more closely at home, take a very small one from an inconspicuous spot.

Sphagnum magellanicum x1

S. contortum, showing how the branches are arranged x1

Carolyn Reider

Sphagnum palustre with the longer branches very dry and pale x1.5

Sphagnum – Peat Moss

Besides being exceptional mosses structurally, members of the genus Sphagnum influence plant communities more than any other mosses. They often completely dominate a wetland habitat and modify it to such an extent that only certain other species can live there.

All species of the genus Sphagnum have several unusual features in common. The gametophytes are very large and highly branched with a vertical main stem which has many branches crowded near the tip of the plant and bunches of branches more widely spaced below. The plants are normally packed tightly together with only the upper parts visible. The leaves, which lack midribs, have large clear water storage cells in a network of narrow green photosynthetic cells. This unique pattern of green and clear cells gives peat mosses their characteristic pale color when dry. In some species of Sphagnum growing in full sun, red non-photosynthetic pigments mask the green pigments. Besides differing somewhat in color, the various species of Sphagnum differ in size, in the number of branches around the stem tip, and in the shape of the leaves. Species identification depends to a great extent on examination of the cells of the leaves and stems under high magnification.

In Sphagnum the sporophyte is elevated on a short side branch of the gametophyte. It consists of a fat round capsule with an operculum that is pushed off by pressure building up inside the drying capsule. There are no peristome teeth and no stalk.

Sphagnum may grow in small patches in moist forests, especially near springs, seeps, or small streams on more acid soils. It is most important ecologically, however, in peat bogs, where it grows in huge masses, especially in northern areas. Because Sphagnum contributes to increased acidity in bogs, decomposition is very slow. The dead Sphagnum remains in bogs for centuries and, in the far north, is sometimes mined for use as a water-holding soil additive or as a fuel.

In peat bogs several species of Sphagnum may be found together, with few other bryophytes present. Of the higher plants that accompany Sphagnum in peat bogs, by far the commonest are members of the heath family, such as Bog Laurel and Labrador-tea, which tolerate low nutrient levels better than most plants.

Tetraphis pellucida gametophyte with mature sporophytes x4

Immature sporophytes x4

Capsules with four teeth x6

Tetraphis in early summer showing gemmae cups x3

Carolyn Reider

Tetraphis in late fall with rust-colored leaves x4

Tetraphis pellucida – Georgia

Tetraphis is a small acrocarpous moss that is most easily recognized when reproducing asexually by making gemmae. The gemmae are located in cup-like areas at the stem tips. The cups are formed by a group of leaves that are larger and rounder than the lower leaves and grow closer together, overlapping at the edges. Tetraphis is also fairly easily identified when it has mature sporophytes. Examine a capsule carefully with very good eyes or a hand lens and, if no operculum is present, it can be seen that the capsule has only four peristome teeth. The teeth are large and fairly stiff and are unlike those found in any other moss. Besides the four peristome teeth and gemmae cups, another feature that may make you suspect that a moss is Tetraphis (even from a distance) is the rusty color of the leaves often found at the tips of older plants.

After sexual reproduction has taken place in late summer or early fall, new sporophytes topped by white hood-like calyptras grow up from the tips of the gametophyte plants. In fall and winter, patches of Tetraphis are often seen in which some plants have gemmae cups, some have old sporophytes from the previous year, and some have new sporophytes. Hood-like calyptras are found in some other mosses, but the four exceptionally large peristome teeth of the sporophytes are unique to Tetraphis and another related genus not found in Pennsylvania.

Tetraphis grows on well-rotted wood or on soil with a lot of humus in it. It is always found in locations that are shady and very moist and is more common on acidic soil.

The genus name Tetraphis refers to the fact that the top of the capsule splits into four parts, the four peristome teeth. The species name is based on the translucent quality of the leaves, which Tetraphis shares with many other mosses. The old genus name, Georgia, would make a good common name. The genus was originally named in honor of George III, who was king of England when the description of this moss was first published.

Diphyscium foliosum with mature and immature sporophytes x2

D. foliosum enlarged to show details of leaves and of sporophytes x4

Jack Selby

Mature sporophytes of Buxbaumia aphylla x6

Diphyscium foliosum – Grain Moss

When its large-capsuled, oddly shaped sporophytes are present, this is one of the most easily identified mosses. The gametophytes have short upright stems with very dark green, tongue-shaped leaves and grow on steep shaded banks where other plants do not compete with them. They often cover an area of a square foot or more. The tips of stems about to undergo sexual reproduction develop narrow leaves with fine hair points, making their total length four or five times longer than the normal leaves by the time the sporophytes are mature.

Because of their odd method of spore release Diphyscium sporophytes do not need stalks to elevate their capsules. Instead, the large, somewhat flattened capsules are constructed so that rain drops hitting their flexible walls make the spores puff upward, often to a height of two inches. Each capsule is about a fifth of an inch long and is slanted, rather than upright, with a small conical operculum which falls off to reveal an upright, pleated conical tip. The spores are puffed out a hole at the very top.

Since Diphyscium foliosum grows on steep shaded banks, all the capsules usually develop with their flattened side turned the same direction to get the best light for photosynthesis. When the sporophytes mature, their color fades to a pale gold which stands out against the dark green of the gametophytes. The color of the sporophytes and their odd shape reminds some people of grains of wheat. Others compare the group of sporophytes to scared rabbits in the grass. D. foliosum grows on soil banks or less often on a thin layer of humus over rock. It is not really common, but is found fairly regularly along trails in deciduous forests.

A cousin of Diphyscium foliosum that is even more unusual is Buxbaumia aphylla. Its capsules are similar to those of D. foliosum in shape. They are raised on short stalks over an inconspicuous mat of filamentous gametophyte, or protonemata. (See Pogonatum pensilvanicum, p. 25.) B. aphylla is found on disturbed soil and is commoner in sandy areas like the Pine Barrens of New Jersey than it is in Pennsylvania. A. J. Grout (see p. 88) suggested the common name Bug-on-a-stick for Buxbaumia.

Rob Fergesen

Atrichum angustatum x2

Atrichum undulatum x1.5

Atrichum – Catharinea

Members of the genus Atrichum are common in Pennsylvania, both in forests and along roadsides. The gametophytes have long narrow leaves that radiate out from the upright stems when moist, but twist and become contorted when dry. The leaves have teeth on their edges which are just visible with an 8x lens in good light. The sporophytes have narrowly cylindrical capsules that are usually somewhat curved. Each capsule is capped by an operculum with a long narrow central projection. The operculum and upper part of the capsule of an immature sporophyte are covered by a delicate smooth green calyptra. When both the calyptra and operculum have fallen from the top of the sporophyte, a pale disk is visible which appears to have a slightly larger diameter than the upper part of the capsule just below it. Thirty-two tiny peristome teeth extend from the upper edge of the capsule to this disk. (They are visible at 10x magnification.) The sporophytes are nearly mature in early fall, release their spores through early winter, and remain present on the gametophytes even in early spring.

Atrichum angustatum is a very common species of the genus which is small with narrow leaves and is found on relatively dry sites such as road banks. This species often covers large areas with thousands of stems packed closely together, especially on mounds of soil around the roots of a fallen tree. In late fall, any disturbance of the ranks of sporophytes sends a cloud of fine greenish brown spores into the air.

Other common members of the genus belong in the broadly defined species Atrichum undulatum (as described by Crum and Anderson in *Mosses of Eastern North America*). Members of the commonest variety of this species usually grow on organic soil in shaded areas of the forest, often along streams. Its broader leaves may appear somewhat wavy on the edges, and the midribs of the leaves are very obvious. The reason they are so easy to see is that each midrib has several ribbons of extra photosynthetic tissue running along it. Although these are not easily visible (even with a hand lens), they do make the midrib stand out and appear bigger than it really is. All species of Atrichum have these ribbons on the midrib, but on A. undulatum they are particularly obvious.

The old scientific name Catharinea comes from Catherine the Great, after whom the genus was originally named.

Polytrichum with male heads x2

Carolyn Reider

Polytrichum ohioense with calyptras covering very young sporophytes in early spring x1.5

Dry Polytrichum capsules with ridges at the angles x3 (See p. 8 for an enlargement showing the teeth.)

Immature sporophytes of P. ohioense in late spring x2 Calyptras cover only the top part of the capsule.

Polytrichum – Hairy-cap Moss

As the common name suggests, the sporophytes of this genus have hairy caps. The calyptras covering the developing capsules are white to pale brown and are composed of fine hairs that separate and become more visible at the lower edge of the capsule. The capsules are round in cross-section when immature, but develop angles when dry. Under the operculum of each capsule is a ring of sixty-four tiny peristome teeth barely visible with a 10x lens. (See the center right photo on p. 8.) The gametophytes also have unusual features. Because the upper surface of the leaf is lined with ribbons of extra photosynthetic tissue, the leaves are opaque and appear to be thicker than most moss leaves. The shape of the leaf is odd, too, with a broad base around the stem, and the upper part often only two-thirds as wide. All these features make the genus Polytrichum easy to identify.

Members of this genus have male and female reproductive organs in separate colonies. The females can be recognized by the sporophytes on top, and the males by the flower-like splash-cups where the sperm-producing organs are found.

There are seven species of Polytrichum in Pennsylvania, of which at least three can be distinguished from the others with only a hand-lens. For the remainder, a compound microscope is needed for definite identification, although good guesses can be made on the basis of easily visible features plus the habitat where the moss is found. A Polytrichum on well-drained banks or soil mounds in a hardwood forest is most often P. ohioense. Its capsules mature in June and are narrow, with the calyptra not reaching the base of the capsule (as in the photo opposite). P. pallidisetum is similar but commoner farther north, in coniferous forests.

The three species easiest to identify with a hand-lens are Polytrichum commune, P. juniperinum, and P. piliferum. P. commune has teeth on the edges of its leaves and grows in open sunny areas in fairly dry to rather moist areas, including wetlands. Its capsules mature in fall and are not much longer than they are wide. Its calyptras are longer than its capsules. P. juniperinum and P. piliferum have short capsules too, but their leaves have smooth edges. In both these species the leaves fold up along the edges so that a narrow strip of the light-colored lower surface shows from the top. (See next page.) The leaves have long sharp

Polytrichum commune growing with P. juniperinum x2

Polytrichum commune
with sporophytes x3

Polytrichum piliferum x3

Pogonatum pensilvanicum
with young sporophytes x4

P. pensilvanicum with
older sporophytes x4

Polytrichum continued

hair-like points at the tip which are brown in P. juniperinum and white in P. piliferum. (Use a lens and good light.) P. juniperinum, which has blue-green or gray-green leaves, is very common in dry open areas. Sometimes P. commune and P. juniperinum grow together in adjacent patches in areas of sun and moderate moisture, with the P. juniperinum in the drier spots. P. piliferum is a smaller hairy-cap moss found in very dry locations, on thin soil over solid rock or on south-facing shaley slopes.

Pogonatum pensilvanicum

This moss presents a very unusual appearance because the sporophytes appear to be growing alone rather than on top of the gametophytes. Actually, the gametophyte stage is present, just in a reduced form. In most mosses when an airborne spore lands on a suitable moist substrate, a microscopic green filament grows from it. After these filaments (also called protonemata) have grown for a while, buds are formed on them which grow into the standard gametophytes with stems and leaves. In Pogonatum pensilvanicum, a large thick mat of protonemata is formed, but no buds are produced until the plant is about to undergo sexual reproduction. The very short upright stem of the gametophyte makes just a few leaves and, above them, the egg- and sperm-producing organs. Thus when the sporophytes resulting from the sexual reproduction have matured, their bases are at ground level on the protonemal mat.

Members of the genus Pogonatum have hairy calyptras and have peristome teeth similar to those of Polytrichum (but usually fewer in number). The main difference from Polytrichum is that when the capsules dry they remain round in cross-section, rather than angled. You can be sure of Pogonatum pensilvancum, though, without waiting for its capsules to dry. It is easy to recognize by its obvious mat of protonemata and by the long narrow drawn out tips of the few leaves. (Another more southern species of Pogonatum with protonemal mats has blunter leaves.)

Pogonatum pensilvanicum grows on bare soil, often on clay banks, and may cover a square foot or more. It does best where steep slopes have allowed for slippage and fresh, moist, bare patches of clay are exposed.

Aulacomnium palustre
with whip-like branches x3

A. palustre with male heads
and one whip-like branch x4

Aulacomnium heterostichum with old sporophytes x5

Aulacomnium

This genus of acrocarpous mosses includes two species that are fairly common in Pennsylvania. Both have leaves with prominent midribs, lots of brown hair-like rhizoids on the stem between the leaves, and curved capsules that have vertical ribs when dry. Also during drying, the capsules shrink unevenly and end up narrower in the upper third than in the lower two-thirds.

Aulacomnium palustre has long narrow leaves that have fairly elongate pointed tips. The gametophyte reproduces asexually by gemmae found at the end of long, bare, whip-like branches. As the stem elongates before producing the gemmae at the tip, it makes progressively smaller and more widely spaced leaves. Some clumps of A. palustre have egg-producing organs and others have sperm-producing organs. The males have many small, dark, sperm-producing organs at the tip of each upright stem, surrounded by a star-like group of extra-large leaves, which makes them resemble the flower-head of a daisy. When the sporophytes produced by this sexual reproduction have finished growing, they are often an inch and a half in height.

The name palustre is from the Latin for marsh, and Aulacomnium palustre certainly lives in wet habitats. Most common in areas of acid soil, it grows in shaded swamps, in seeps dominated by skunk cabbage, in brush along streams, and in bogs.

If found without the characteristic capsules of Aulacomnium, A. heterostichum might at first glance be mistaken for a Mnium (see p. 41). When examined carefully with a lens, it is clear that the leaves do not have the specialized border found in Mnium and the teeth on the upper half of the edge of the leaf are very large and much more irregularly arranged than in Mnium. The other unusual feature of the leaves of A. heterostichum is that they are mostly curved in one direction. The stems are often somewhat curved as well, leaning in one direction over most of the colony.

Aulacomnium heterostichum grows most often on steep, rather dry banks in broad-leaved forests, where it may be associated with Bartramia pomiformis (see p. 29). The bare soil exposed by erosion of these steep slopes favors the growth of both these species. A. heterostichum may also be found in moist woods, usually at the base of a tree, where the drainage is better than on the forest floor.

Apple Moss with immature sporophytes in early spring x1

Apple moss with slightly older sporophytes x2

Sporophytes showing the wrinkles of maturity x3

Bartramia pomiformis – Apple Moss

The common name of this moss comes from the shape of the capsule of the sporophyte, which is spherical, rather than cylindrical as in most mosses. The gametophytes are found in tightly packed clumps, with the lower part of the stems covered by brown hair-like rhizoids. The long narrow leaves are bright green with a dull surface. Each leaf is about a fifth of an inch long and appears to be somewhat wavy or twisted, especially when dry. The leaves tend to curve and lean downward.

Sexual reproduction occurs in late fall, and half-inch tall immature sporophytes are present by late winter. When an early spring rain and warmer temperatures break the drought of winter, the straight narrow immature sporophytes suddenly swell at the tip to form the capsules. During maturation each capsule is at first smooth, shiny, and round like a green apple. Then it gradually turns brown and becomes wrinkled when mature and dry. Under certain conditions of light and moisture it may even go through a red stage as it develops.

When lacking sporophytes Bartramia may be confused with Dicranum scoparium (which may be found in similar habitats), since the leaf shape and the overall aspect of the clumps of both mosses are similar. The leaves of Bartramia, however, are not shiny like D. scoparium leaves. Also, Bartramia leaves are not as regularly curved or as consistently turned in one direction as those of D. scoparium.

Bartramia is not a common moss, but it is well worth looking for. It grows on moist rock outcrops near waterfalls or on mineral soil of steep shaded banks, where it may be found with Aulacomnium heterostichum (see p. 27). Both are most often found on acidic substrates. Another species of Bartramia, B. ithyphylla, is sometimes found on cliffs in the mountains. Its leaves are straight rather than wavy when dry, and are very pale from the base about a quarter of the way up.

The Apple Moss genus was named for John Bartram of Philadelphia, the first Pennsylvania botanist. Along with other plant specimens, Bartram collected American mosses and sent them to Europe in the mid-eighteenth century to be described and named. The species name pomiformis means apple-shaped in Latin.

Moist Bryum argenteum x4

Bryum argenteum
with sporophytes x6

Very dry B. argenteum x3

Funaria capsules x3

Funaria hygrometrica x1.5

Bryum argenteum – Silver Moss

This tiny moss is extremely common on disturbed soil or in dry rocky areas. It is recognizable by the size and color of the plants, which appear to be a silver green because of the presence of dead empty cells near the tips of the leaves. Since the leaves fit tightly together around the stem, the white leaf tips are often more obvious than the greener base of the leaf. Younger fast-growing plants are somewhat greener; and older, dry, slow-growing plants may be almost gray. The long-stalked reddish sporophytes of Bryum argenteum only occur in small numbers compared to the area covered by the gametophytes. The drooping capsules are similar in shape but not size to those of Mnium and of Rhodobryum. (See pp. 41 and 45.) The sporophytes mature in late fall.

Bryum argenteum is common in cities and along sandy paths in forests. It grows in the cracks in pavement and between bricks in garden walks, and is generally found wherever human beings settle or travel. The tips of its stems are easily knocked loose and carried on passing feet to grow in new locations.

Funaria hygrometrica – Cord Moss or Cinderella

This small acrocarpous moss is also associated with people, but it may grow in either dry or damp places. Its leaves are bright grass-green and crowded together near the tops of the stems. The gametophytes are rather ordinary and resemble those of other genera such as Bryum (in which most species have similar green leaves). Funaria hygrometrica stands out because it undergoes sexual reproduction very freely and produces huge numbers of tall sporophytes, up to two inches in height. Their stalks are very delicate and inclined to twist together like the strands of a rope or cord. The capsules are pear-shaped but asymmetrical, with oblique mouths. They are heavy enough to make the stalks bend. When the capsule is full-sized but still green, the operculum is yellow with a red-brown rim. The old capsules are very wrinkled.

Funaria hygrometrica grows on the bare ground after forest fires or at camp-fire circles a year or two after they are abandoned, giving it the common name Cinderella. In more populated areas, though F. hygrometrica prefers damp soil in greenhouses, it also is found in drier spots, even growing between sidewalk blocks. The huge numbers of spores it produces lie in the soil ready to grow whenever the right amount of light and moisture is present.

Dicranum scoparium with mature sporophytes x1.5

Carolyn Reider

Dicranum scoparium x1

Dicranum scoparium – Broom Moss

This species of Dicranum grows in large clumps with the leaves all turned in one direction. The leaves are a shiny grass green. Each leaf is very long and narrow with the widest part near the base. It is about half an inch long and curved from base to tip in a continuous arc. All the leaves in the clump have parallel curves to their leaves, giving the clump a very orderly appearance. The stems are covered with brown, intertwined, hair-like rhizoids.

Male and female reproductive organs are found on separate gametophyte plants in Dicranum. Some male plants in some species of Dicranum (including D. scoparium) are dwarves which grow on the hair on the stems of the female plants. (Thus the sperm don't have far to swim.) After sexual reproduction has taken place, each sporophyte grows to an inch or more in height, including the curved cylindrical capsule. There is a long, drawn-out tip on the operculum. If sporophytes are present, each upright plant has only one (or very rarely two), attached to the older branches. As seen in the photograph on the facing page, new upright plants without sporophytes may also be present.

Another good common name for Dicranum scoparium is Wind-swept Moss, because the leaves all appear to be blown by a strong breeze. Broom Moss, the old common name, may seem strange to modern readers, but illustrations of old-fashioned brooms often show straw or twigs bent from heavy use. Indeed, the name scoparium refers to a broom.

Dicranum scoparium is very common on drier forested slopes as well as in moist areas near woodland streams. It may grow on soil, on rock, or even on rotted wood, and seems to be found more often in areas of somewhat acid soil. It is often found with Leucobryum and Polytrichum (see pp. 39 and 23).

The genus Dicranum includes a large number of species. Besides D. scoparium only one other local species in the genus has shiny green leaves. D. polysetum is not as common as D. scoparium, and is larger, with more than one sporophyte at the tip of each upright branch of the gametophyte. Its leaves have wavy margins and are not all curved in one direction. D. polysetum is found in forests with a mixture of hemlock and hardwoods.

Dicranella heteromalla in early fall x2

Curved leaves of D. heteromalla x8

D. heteromalla in mid-winter with mature sporophytes x3

D. heteromalla old sporophyte with oblique mouth x6

Dicranella heteromalla

This very delicate small moss resembles a miniature Dicranum. (See p. 33.) The shiny, green or yellow-green leaves are extremely narrow and curved. They tend to all lean in the same direction, as in the upper right photo on the facing page. Each leaf is less than a tenth of an inch long and each upright stem rarely taller than half an inch. When Dicranella heteromalla is growing vigorously it often appears as a low mound a couple of inches across with the stems tightly packed together.

After sexual reproduction has taken place during the summer, the young sporophytes of Dicranella heteromalla grow up to about half an inch in height and start to expand their capsules. The stalks and capsules of the sporophytes turn from green to yellowish brown during the fall. If you find a mature capsule in late fall or winter when the calyptra has fallen off but the operculum is still present, you will see that its "nose" is as long as the capsule and bent at an angle to it. In early winter, after the spores are released, the capsule continues to dry out and develops wrinkles running not quite lengthwise along it. Because of this uneven wrinkling, by the end of the winter the old capsule shows the oblique mouth that is characteristic of the species.

Dicranella heteromalla grows on bare mineral soil. It is very common on shaded banks and tree-fall mounds and along trails in forests. On a suitable substrate, a group of D. heteromalla clumps often covers an area of a square foot or more. D. heteromalla is often found near Atrichum angustatum (see p. 21) on tree-fall mounds. Both these species help hold the soil and add organic material to it, making the mound more suitable for the growth of higher plants.

A small moss on banks or tree-fall mounds with very narrow, curved leaves and old capsules with oblique mouths is almost certainly Dicranella heteromalla. The capsules of its sporophytes are three times as long as wide and their stalks are yellow-brown. The other fairly common species of Dicranella in this area have red-brown stalks and shorter capsules that lack the oblique mouths found in D. heteromalla.

Carolyn Reider

Fissidens cristatus on a tree base in a moist forest x3

Dick Heller

Fissidens bushii
from disturbed soil x3

Fissidens adianthoides
from rocks in a stream x2

Fissidens

This is a genus of small, upright, seldom-branched mosses which have a peculiar and easily recognized arrangement of leaves. The gametophytes generally grow from a quarter inch to an inch in height with all the leaves in one plane, forming two rows on either side of the stem. The basal part of the lower side of each narrow leaf appears to be overlapped by the leaf below it. Actually it fits into a pocket in the basal part of the upper side of that leaf. (The rest of the leaf has the normal one layer of cells, but the area above the midrib near the base of the leaf has two separate layers like a long shallow pocket to enclose the base of the leaf above. The leaves thus fit together much the way the leaves of the typical garden Iris do.) The sporophytes are rarely over half an inch tall, with cylindrical capsules that are proportionally small (often less than a twentieth of an inch in length). The capsules may be straight or somewhat curved. In some species the sporophytes are at the tips of the gametophytes, as expected in acrocarpous mosses, while in others they are lower on the stem. The tiny plants of Fissidens usually grow in fairly large clumps and often seem like a very small garden of delicate little ornamental plants.

About ten species of Fissidens have been found in Pennsylvania, but only four of them are more than a third of an inch in height, which means they are often so small they are not noticed. Most Fissidens species are found in moist forested areas in shaded spots by streams or seeps, sometimes on soil or rock, and sometimes on bark at the base of a tree. One of the larger species grows on disturbed soil on road-banks and in garden beds. An unusual and rather uncommon species (F. fontanus) is aquatic and may grow just below the water surface in streams or swamps or at great depths in clear lakes. Its stems are up to four or five inches long, lined with two rows of long, narrow, widely spaced leaves.

Most species of Fissidens are rather similar in appearance and require the use of a compound microscope for definite identification. For example, although F. adianthoides is generally larger than F. cristatus, their size ranges overlap, and the only distinct differences are in the size and arrangement of the leaf cells. At least, with the unusual leaf features of Fissidens, you can be sure that you have the right genus.

Dick Heller

Sporophytes x2

Carolyn Reider

A large cushion of <u>Leucobryum</u> x1

Naja Kraus

A small cushion of <u>Leucobryum</u>
with rather large thick leaves x1

Leucobryum – White Moss or Cushion Moss

The two species of *Leucobryum* found in Pennsylvania differ significantly from other mosses in having leaves made up of several layers of cells, rather than the usual leaves with one layer of cells. In *Leucobryum* the outer layers of cells are dead and have no chlorophyll. These cells are water-filled and fairly translucent when the leaves are damp and air-filled and less translucent when dry. This is why the leaves of *Leucobryum* appear to be whitish or pale green if dry and a stronger green if wet. Because the cell layers are arranged with the living green photosynthetic cells covered by dead cells, the living cells are better protected from dehydration than those of most moss leaves. Both species of *Leucobryum* live primarily in drier areas of the forest. Their growth habit with the small upright plants tightly packed into a cushion-like colony probably also helps to slow their loss of water to the dry air above them. The cushions are often two or three inches in diameter and sometimes much larger.

The two species of *Leucobryum* found in Pennsylvania, *L. albidum* and *L. glaucum*, are separated from each other by technical microscopic details of their leaf structure. Some sources suggest that *L. glaucum* has larger leaves than *L. albidum*, but there is a broad range of overlap in leaf size between the two species. The size of the leaves often varies a great deal in neighboring cushions. In some locations it appears that larger older colonies have smaller leaves, more tightly packed together.

It seems as if sexual reproduction does not take place very often in *Leucobryum*, since sporophytes are absent on so many of the cushions. This is even true with conditions ideal for sexual reproduction, after a wet summer when a few cushions have many sporophytes. In *Leucobryum glaucum* asexual reproduction occurs by fragmentation of the plants. Turkeys or other animals that scratch the ground in search of food scatter leaves or other fragments of the moss that can grow into a new plant and start a cushion as more upright plants form around the original one.

Though they may also be found in damper areas, both species of Leucobryum are often found on well-drained slopes above streams, accompanied by *Polytrichum ohioense* (see p. 23) and by *Dicranum scoparium* (see p. 33).

Upright and horizontal stems of Mnium affine x1.5

Male plants of Mnium affine x3

M. cuspidatum with narrow immature sporophytes x2

Carolyn Reider

Mnium with swelling capsules x1

Mnium mature capsules x2

Mnium

This genus of acrocarpous mosses includes several species that are especially noticeable, either because they are common or because they are large. All have gametophytes with broad, green translucent leaves with specialized margins. The capsule of the sporophyte is not usually upright, but leaning or hanging at the top of the stalk. Mnium is a large genus including about ten species in Pennsylvania. Four of the most outstanding species are described on this page and the next.

Mnium affine and Mnium cuspidatum are fairly similar. Both have upright stems when reproducing sexually, but leaning or horizontal stems when not reproducing. Their stems may have some brown hair at the base; their leaves are pointed, toothed on the edge, and about twice as long as wide. In both species, the teeth on the leaves are large and arranged singly at intervals along the margin. The upright stems differ in height, with M. affine often two or three inches tall and M. cuspidatum usually less than two inches. Since the range of heights overlaps, use a hand-lens to check the teeth on the edges of the leaves. In M. affine the teeth are arranged along the whole margin of each leaf, while in M. cuspidatum there are teeth only on the half of the margin towards the tip of the leaf. Mnium affine has female and male reproductive organs on separate plants, with the sperm-producing organs in a black mass at the tip of the male plants. In M. cuspidatum both types of reproductive organs are found on the same plant.

Mnium cuspidatum is very common, growing both in forests and in disturbed habitats like shady lawns (see p. 12) and edges of paths. It is especially common on low soil mounds in deciduous forests. In late fall it makes huge numbers of tall, narrow, immature sporophytes, which swell and droop at the tip, forming capsules in early spring. On the edges of the colony, there are often areas of low, creeping, non-reproducing stems with the leaves arranged in two rows. The commonest type of M. affine, with large teeth as described above (M. affine var. ciliare), grows in moist forests, in less disturbed areas than M. cuspidatum. It grows in deciduous forests near streams, with upright stems in the center of the colonies and some leaning or horizontal ones on the edges. Its reproduction occurs in summer; the sporophytes mature in early spring and release their spores in May, as in M. cuspidatum.

Mnium hornum
in late spring x2

M. hornum
in early fall x2

Mnium punctatum
in early summer x2

M. punctatum in winter x1.5

Mnium hornum has narrower leaves than most species in the genus, usually four times as long as broad. On the margins of the leaves there are regularly spaced teeth, which are double, (though this feature is not visible with a hand-lens). The stems are often an inch or two long, not upright but leaning, with the adjacent stems all leaning the same way. M. hornum has male reproductive organs at the tip of a plant in the center of a ring of much larger leaves than the others on the stem. It seems to produce sporophytes only rarely. If you are lucky enough to find them, they should be one to two inches tall with a very thin stalk, a drooping capsule, and an operculum without a long "nose" on top. Mnium hornum is very common on banks of streams in the woods, especially in areas of acid soil. It often makes uniform mats covering large areas of ground exclusively, not mixed with other mosses. Its narrow, toothed leaves and its habit of leaning or hanging in mats on stream banks make this Mnium species recognizable.

Mnium punctatum is a very large moss with leaves nearly round, or at least less than twice as long as wide. Each leaf has a narrow dark margin and a prominent dark midrib that extends all the way to the margin at the end of the leaf. The leaf cells are very large, easily visible at 8x magnification. As the leaves age, they change in color from a pale yellowish green in early spring to the characteristic dark translucent green in late summer, fall, and winter. The leaves may be half an inch long, and the stems are often between one and two inches in height. The stems are covered with matted hair-like rhizoids which are dark brown except on new growth. M. punctatum has male and female reproductive organs on separate plants. The male plants have the sperm-producing organs concentrated in a round black area in the center of a flower-like head. (See p. 40 for a picture of male M. affine, which has similar male heads .) As the sporophytes mature their capsules expand and then droop as expected in Mnium. Each operculum has a long straight "nose" on top. M. punctatum grows in very wet areas in seeps or along the edges of spring-fed streams.

Some authors put Mnium punctatum in the genus Rhizomnium, M. affine and M. cuspidatum in the genus Plagiomnium, leaving only the double-toothed species like M. hornum in the genus Mnium.

Rhodobryum roseum
with mature sporophytes x2

Naja Kraus

Rhodobryum roseum showing black centers
where sperm-producing organs are located x1

Rhodobryum roseum – Rose moss

This large moss is particularly striking because of the flower-like arrangement of the leaves. The upright stems of _Rhodobryum_ are nearly bare in the lower half, with the leaves crowded together in a rosette around the tip of the stem. Each leaf is usually more than a quarter of an inch long, narrow at the base and broad towards the tip, with an abrupt sharp point. The upright stems, which are up to two inches tall, are connected underground by creeping horizontal stems to form patches often over a foot across.

In late summer or early fall each patch normally produces either sperm or eggs. At that time the male plants show black in the centers of the rosettes where large numbers of sperm-producing organs are located. When fertilization is successful, the female plants often develop more than one sporophyte on each upright stem (in contrast to most mosses where only one is found per stem tip). The sporophytes are one to one and a half inches tall with an inclined or nodding capsule. Unfortunately, they are produced only in very moist habitats. With few spores released, the establishment of new clumps of _Rhodobryum_ is uncommon. The large patches observed in relatively undisturbed forests are probably quite old.

Rhodobryum grows in moist forested areas in smaller clumps at the base of trees or in larger ones on organic soil over rock. It may grow either at acidic sites or at sites with calcareous rock.

The nodding capsules seen in _Rhodobryum_ are also found in _Bryum_ (see p. 31) and in _Mnium_ (see p. 43). The large size of _Rhodobryum_ and its flat rose-like array of leaves distinguish it from _Bryum_. The male plants of _Rhodobryum_ and of some species of _Mnium_ might be confused, except that the lower part of the stem is bare in _Rhodobryum_ and leafy in _Mnium_.

Some experts split this species into two and have concluded that the plants found in eastern North America should be called _Rhodobryum ontariense_. The differences between the two species are slight, however, and the older name is easier and more descriptive.

Anomodon attenuatus, damp, in late fall x2

A. attenuatus, dry, in summer x1.5

A cushion of Bryoandersonia x1

Detail of leaves x4

Carolyn Reider

Anomodon attenuatus

The drooping finely drawn-out ends of the main branches of the gametophyte make this species of Anomodon recognizable. In common with other species of the genus, its leaves are dull and yellowish green. They have short pointed tips, not long, drawn-out ones. (The leaf shape, together with the attenuate branch ends, distinguishes A. attenuatus from A. rostratus, which is also very common in Pennsylvania.) The sporophytes of Anomodon are between half an inch and an inch in height. Their capsules have a narrow upright cylindrical shape and are only a tenth of an inch tall. The sporophytes are only found at the dampest sites.

Anomodon attenuatus is most often found on bark at the base of trees. (For this reason some people call it Tree-skirt Moss.) Extensive mats of it can also be found on soil or rocks. This moss is extremely common in areas where the soils are formed from basic rocks, but may also be found in acidic areas. It is often seen growing near Spoon Moss.

Bryoandersonia illicebra – Spoon Moss

This species is characterized by very crowded, tightly overlapping, wide, shiny leaves, often golden in color. Each leaf is nearly as broad as it is long and deeply concave. The common name is derived from the wide concave shape of the leaf, like the bowl of a spoon. The stems of Bryoandersonia are highly branched with the ultimate branches growing more or less upright and tightly crowded together so that the clump is often cushion-like. The sporophytes of Bryoandersonia are produced in fall, but only rarely. If you find them, you will see that the capsules are cylindrical but bent, and each operculum has a long "nose".

Bryoandersonia is usually found on soil in very moist forested locations. It is more common in areas of neutral or basic soil than in acidic areas. Often Bryoandersonia is found on the ground around the roots of a tree where Anomodon attenuatus grows in a band around the lower part of the trunk.

This genus was named after Lewis Anderson, who is one of the authors of *Mosses of Eastern North America*. (See p. 88.) It is distinguished from the large difficult genus Brachythecium by its broad, deeply concave, and tightly crowded leaves.

Rob Fergesen

Tree-Moss turf in winter, showing the tops of the "trees" x1.5

Dick Heller

Horizontal underground stem with very new growth at left and old "trees" at right x1

An old "tree" with several young ones around it x1.2

Climacium – Tree-Moss

Climacium is the largest pleurocarpous moss that is common in Pennsylvania. It often makes patches of several square yards in low wet woods. Its horizontal underground stem forms upright stem branches that emerge from the ground. Then about half an inch above the surface these upright stems branch repeatedly, taking on the form of small trees. The bare lower part of the upright stems and the hidden horizontal stem linking the small "trees" are key characteristics in recognizing this moss. Also the shape of the leaf is important: broadest near the base and gradually narrowing towards the tip.

As the underground stems grow they produce new upright branches behind the growing tips. When *Climacium* has favorable conditions it forms so many upright branches so close together that it is like turf, and you may have difficulty seeing the bare lower part of the upright stems without picking one of them.

The sporophytes of *Climacium* are not seen as often as those of many other mosses. When they do occur, several sporophytes may be found scattered along an upright branch of the gametophyte. Each sporophyte has a cylindrical capsule atop a stalk that is about an inch and a half in height.

There is another moss with a similar underground stem and with the lower part of the upright stems bare (see *Rhodobryum*, p. 45). Tree-Moss, however, branches extensively above ground and has leaves broadest near the base, rather than near the tip as in *Rhodobryum*.

Large amounts of Tree-Moss are most likely to be found on soils derived from neutral or basic rock, such as limestone. Tree-Moss is often found with *Bryoandersonia* (see p.47). It flourishes with plenty of moisture and at least light shade, often in wooded wetlands or even in damp lawns.

There are two species of *Climacium* in North America. *C. americanum* is common in Pennsylvania, while *C. dendroides* is more often found farther north in wet areas surrounding bogs. The most reliable difference between the two is in microscopic details of the leaves. However, if you find straight-sided cylindrical capsules an eighth of an inch or more long, the tree moss is *C. americanum*. *C. dendroides* has shorter, more rounded capsules.

Entodon cladorrhizans dry x2

Damp E. cladorrhizans with maturing sporophytes x4.5

Entodon seductrix x4.5

E. seductrix with old sporophytes x4

Entodon

A pleurocarpous moss with shiny, concave leaves and straight, cylindrical capsules held vertically is very likely to be a member of the genus Entodon. There are two common species of Entodon in Pennsylvania.

Entodon cladorrhizans has shiny, yellow-green leaves on stems that creep horizontally very close to the substrate, which is most often rotten wood. The stems are fairly regularly branched and completely covered with very regularly arranged leaves. Each stem is flattened in such a way that the leaves on its top and bottom cling tightly to it, while the ones on the sides extend out from it. The oval leaves are concave and pointed. They are attached very close together, overlapping a lot with their neighbors.

After sexual reproduction has taken place in Entodon cladorrhizans, the mature sporophytes appear in late fall or in winter. The capsules and their stalks range from yellow to reddish brown. The operculum is cone-shaped, but a little crooked.

Entodon cladorrhizans is quite common on rotten logs. It may grow down on the soil in relatively dry woods, but in moist woods is found up on logs, rocks, or tree bases.

In Entodon seductrix the stem branches grow upward from a horizontal main stem, and are closely covered with leaves extending out equally in all directions. Thus, the cross-section of the stem with its leaves is round rather then flattened as in E. cladorrhizans. E. seductrix leaves are elliptical and less than twice as long as broad, with only a small point at the outer end. The sporophytes of E. seductrix are very similar to those of E. cladorrhizans.

Of the mosses in this guide, the gametophytes of Entodon seductrix most closely resemble those of Bryoandersonia (see p. 47), which also has shiny, broad, concave leaves arranged in much the same way. The upright stems of E. seductrix are narrower and not as crowded together as in Bryoandersonia, and its leaves are narrower and usually greener. E. seductrix grows in drier areas and more often on wood or rock than on soil, while Bryoandersonia normally lives on moist soil. With the sporophytes present the distinction is easy, since E. seductrix has straight capsules and Bryoandersonia does not.

Carolyn Reider

Fontinalis stem tips showing the shape of the leaves x1

Bases of a group of Fontinalis plants attached to a rock x1

A Fontinalis stem tip and three points of attachment x1.5

Fontinalis

This genus includes at least half a dozen species of large aquatic mosses in Pennsylvania. They are commonly found in fast-flowing rocky streams in forested areas. Each plant is connected to its substrate (normally a rock) by a single point at the base of the black, wiry, much-branched stem – not at many points along a creeping stem as in other mosses attached to rocks in streams. The older part of the stem at the base is often leafless for a few inches. At the tips of the branches the leaves are closely overlapping in three vertical rows. The leaves are straight, two to three times as long as wide, and either parallel to the stem or somewhat diverging from it. These leaf features separate Fontinalis from similar but much less common aquatic mosses whose leaves are curved or have long drawn out tips. (For a description of one of them, see Fissidens fontanus, p. 37.)

Because the male and female reproductive organs are on separate plants in Fontinalis, only some of the plants produce sporophytes. Also, compared to their gametophytes, the sporophytes of Fontinalis are quite inconspicuous. Having very short stalks, their capsules are mostly covered by nearby leaves.

When water levels are high in the spring, Fontinalis plants grow quickly, elongating at the stem tips and adding new leaves and branches. A plant may reach eighteen inches in length and is usually at least six inches long. When water flow is less and the plants dry out, they often lose their leaves and stem tips and have to start growing again near the base.

Fontinalis thus is most successful in streams with a constant flow. Streams where conditions are favorable for this species often support many large clumps six or eight inches across near the base, narrowing to the far end where the delicate branch tips flow down with the current.

There are at least seven species of Fontinalis in Pennsylvania. They differ from each other in the details of the shape and arrangement of the leaves. All of the photos at the left show F. sphagnifolia. Other bryophytes found submerged in flowing water include the liverwort Porella (see p. 81), which has broad round leaves, and several kinds of pleurocarpous mosses which are attached to the substrate at many points along their stems.

Hedwigia with moist leaves x2

Hedwigia with dry leaves showing white tips x3.5

Hedwigia with mature sporophytes x3.5

Hedwigia with an old sporophyte x3.5

Orthotrichum anomalum (left) with old sporophytes (red arrows) and young ones under calyptras (white); also Grimmia (right) x4

Hedwigia ciliata

An irregularly branched dull gray-green moss growing on large rocks is most likely to be Hedwigia ciliata. If you check with a lens and find that the leaves have white tips, often drawn out to long, fine points, then the identification is even more certain. The final and completely definitive step is to find the sporophytes, which have almost no stalk. They are so short that they are nearly hidden among the leaves. The capsule is almost round with a barely convex operculum. No peristome teeth are present. The female reproductive organs, and thus also the sporophytes, are located on short side branches with tight clusters of specialized leaves around them, which make this very short branch look somewhat like a flower bud.

Hedwigia lives in more exposed places than most mosses, on rocks or cliffs in open woods. When dry its leaves are tightly pressed against the stems, and when moist they expand and make a wide angle with the stem. Also the color of the leaves changes from gray-green to yellow-green as they absorb water. There is only one species of Hedwigia, though it grows on rocks all over the world and shows quite a lot of variation in the length of the white tips on the leaves. It may be found on rocks of different types, but is more frequent on acidic ones, such as sandstone.

Other species of mosses are often found on the same rocks as Hedwigia. One of them is Orthotrichum anomalum, which also has sporophytes with short stalks and a branching pattern like that of Hedwigia. You will not confuse Hedwigia and O. anomalum, since the latter has leaves so dark as to be nearly black when dry and dark green when wet. Also, all Orthotrichum species have capsules that are cylindrical rather than round as in Hedwigia. Orthotrichum's eight or sixteen peristome teeth are short and broad, sometimes making a star-like pattern at the top of the old capsules, as in the lower photo. Hedwigia has no teeth at all. Other species of Orthotrichum grow mostly on tree bark. They have similar cylindrical capsules and the same type of peristome teeth, but the stalks of their sporophytes are usually even shorter than in O. anomalum. A second rock-loving species of Orthotrichum, O. strangulatum, also has shorter stalks. All have large light-colored, hood-like calyptras with long hairs on them. Another very dark but less branched moss on rocks is Grimmia alpicola.

A mat of Hylocomium showing the fine ultimate branches with leaves indistinguishable x1.5

This year's new growth above the "steps" of the two previous years x1.5

New growth expanding above an old mat x1.5

Hylocomium splendens – Stair-step Moss

Stair-step Moss gets its name from its unusual growth pattern. The main part of each year's growth is a flat lacey spray of tiny branches. The next year's growth arises as a large branch from the center of the spray, growing vertically and then turning to a horizontal position before it expands into a similar network of delicate branches. At an early stage (as in the lower right photo) the new branch is thicker and simpler than the older one below it, but it soon spreads into its lacey pattern. Because each new branch starts about halfway down the last, several years' growth makes a pattern like a flight of stairs. Each step of the stairs is about an inch by an inch and a quarter, and often rises a half inch or more above the one below. In an old mat of Stair-step Moss, besides one or two upper green steps, three or four lower steps may be found, brown but still intact. The branches are covered with tiny yellowish green leaves, clearly visible at 8x magnification.

. The sporophytes of Hylocomium are tall (up to an inch and a quarter in height) with short inclined capsules. They are rarely produced, but look for them in the fall.

Stair-step Moss lives in cool places, in northern areas or in the mountains. It is often found in coniferous forests, where it forms extensive mats along with Pleurozium schreberi. (See p. 61.) In the upper picture on the opposite page, a little Pleurozium is present growing up through the mat of Hylocomium. In Pennsylvania, Stair-step Moss is found in cool ravines in the north or in highlands farther south. Another less used common name for Hylocomium is Mountain Fern Moss, because it is frequently found in montane habitats. The pattern of smaller branches in Hylocomium and its tiny leaves make it seem similar to Thuidium (which is always called Fern Moss, see p. 65), but the major branches like stair-steps are found only in Hylocomium.

The first syllable of the genus name Hylocomium is from the Greek word for forest, while the other root of the name means a revel or celebration – an appropriate name for a moss that revels in northern forests.

Hypnum imponens in summer x1

Carolyn Reider

Hypnum in winter, most likely H. imponens x2

A Hypnum with strongly curved capsules x1.5

Hypnum imponens with barely curved capsules x2

Hypnum

This is a large genus of pleurocarpous mosses whose sporophytes have curved or leaning capsules. Hypnum gametophytes usually are highly branched and grow flat on their substrate, often forming large mats. Their leaves are sharp-pointed and arranged close together on the stem. Since Hypnum lacks midribs, a pleurocarpous moss with easily visible midribs in its leaves can be eliminated as a member of the genus. (Many of these belong in the large, hard to recognize genera Amblystegium and Brachythecium.) Not every pleurocarpous moss with a curved capsule and no midribs is a Hypnum. For many specimens careful examination of the leaf cells and capsules with a compound microscope is necessary to be certain of arriving at even the right genus name.

However, there are several species of Hypnum that can be definitely identified as members of the genus using little or no magnification. They include at least two that are common in Pennsylvania, H. imponens and H. curvifolium. These two species (and two others that are not as common) have leaves that are tightly curved so their sharp points are tucked under and not easily seen. Because of the precise arrangement of the leaves covering them, the stems have a braided appearance, making them easy to spot with sharp eyes alone or with just a little magnification. No members of similar genera of pleurocarpous mosses have leaves arranged this way.

The gametophytes of Hypnum imponens are rather regularly branched, with reddish brown stems. Its sporophyte capsule is somewhat curved to almost straight, and its operculum is shaped like a tall cone (see p. 4, lower right). In contrast, H. curvifolium has a shorter cone-shaped operculum. Its capsules are more curved and often bent nearly horizontal. When dry, they are not smooth but ridged and grooved along their length. The stems of the gametophytes are not as red as in H. imponens. It's not easy to distinguish between these two species. Luckily, their features allow you to be sure they are species of Hypnum rather than of some other genus.

Both Hypnum curvifolium and H. imponens are found on logs, soil, and rock in moist to somewhat drier shaded areas. The color of their leaves is bright green in summer and early fall and a duller green in winter and early spring. Their sporophytes mature in fall.

Pleurozium with one stem scraped to show the color x1

Jerry Hassinger

Pleurozium in winter with old pale leaves x2

Ptilium crista-castrensis x1.5

Pleurozium schreberi – Big Red Stem

This large pleurocarpous moss has pale green to yellow-green leaves covering the reddish brown stems, whose main branches grow nearly upright. The branching of the stems is fairly regular, and they are usually three to six inches long, forming extensive mats. If you see a very large, pale green, mat-forming moss and scrape some of the leaves from the stem with your fingernail to be sure it is reddish, then you are well on your way to identifying Pleurozium schreberi. Look at the leaves more closely to be sure they are very concave and rather broad (often less than twice as long as wide), without much of a point at the tip.

The other important factor in recognizing Pleurozium is its habitat. It grows on acidic soil, often under conifers. Farther north its mats may be several inches thick and cover acres of ground in fairly open forests. In Pennsylvania large but thinner mats of Pleurozium are occasionally found in grassy areas on the edge of woods. Since Pleurozium has female reproductive organs on one mat and male on another, it has been suggested that the large size of the mats may make it difficult for fertilization to take place. If so, this would account for the infrequency of Pleurozium sporophytes. Look for them in the fall, with stalks an inch or more tall and short curved capsules bent nearly horizontal.

Ptilium crista-castrensis – Plume Moss

Its large size and feather-like branching make Plume Moss stand out from similar pleurocarpous mosses. The branches are almost all the same size and are arranged parallel to each other, with a gradual reduction in length towards the tip of the elongate stem. (See also the photo on the next page.) The leaves of Ptilium are a shiny yellow-green with long drawn-out tips which are tightly curved. All the leaves on a branch curve to one side. Because of the tightly curved leaves, both the main stem and the branches are hooked at the tip. The main stem is often several inches long, with the older, lower part of each stem hidden by the new more upright growth of its neighbors. Sometimes the main stem forks, forming two equal stems.

After sexual reproduction takes place in Ptilium, red-brown sporophytes an inch and a half tall are produced near the base of

Ptilium crista-castrensis and a bit of Pleurozium x1

Rhytidiadelphus triquetrus with typical spreading leaves x1.5

Rhytidium rugosum x1.5

R. rugosum, showing down-curved leaf tips x3

Ptilium crista-castrensis continued

the longer branches. The strongly curved capsules mature in fall and release their spores in early winter. Ptilium is found in the northern half of Pennsylvania in cool forests, often in the mountains. As a substrate it prefers old rotten logs or soil rich in humus.

Another distinctive moss characteristic of the northern part of Pennsylvania is Rhytidiadelphus triquetrus (see upper right photo). This large, striking moss with a tousled or shaggy appearance is found on moist, north-facing slopes in northeastern Pennsylvania. It grows with Pleurozium schreberi (see the previous page) and Hylocomium splendens (see p. 57).

Rhytidium rugosum

A large, silky, yellow-green to yellow-brown moss that grows on exposed calcareous rock is likely to be Rhytidium rugosum. The branches (up to three and a half inches long) are thickly covered with leaves that have a broad base and a narrow, curved, hair-like tip. Since the leaf tips all curve downward, the upper side of the branch is smooth and the lower is roughened by the extended leaf tips. The end of each branch appears to be hooked because of the curved leaves clustered there. If examined closely, the leaves clearly have several folds running lengthwise along their lower half parallel to the midrib. They also have wrinkles running perpendicular to the midrib, which are particularly obvious on dry leaves. The name rugosum means wrinkled in Latin.

Sporophytes, very rarely seen on Rhytidium rugosum, are small compared to the size of the gametophytes. The stalks are about an inch tall and the capsules about a tenth of an inch long. The frequent lack of sporophytes may be due to the habitat, which is rarely wet enough for fertilization to occur.

Rhytidium rugosum lives in dry sunny locations on cliff tops, on rocky slopes, and on the walls of abandoned quarries. In Pennsylvania, all known sites for this species are on limestone or other calcareous rock. This substrate specialization helps account for the limited distribution of Rhytidium. It is often accompanied by Orthotrichum species (see p. 55) and sometimes by Bryoandersonia (see p. 47) in shady, damp cracks between the rocks.

Rob Fergesen

Thuidium delicatulum in winter x1.5

T. delicatulum in late summer x2

T. delicatulum in late fall with mature sporophytes x2

Thuidium – Fern Moss

The gametophytes of species of Thuidium have a very precise fern-like branching pattern which is all the more obvious because their leaves are so small as to be nearly invisible, appearing only as a slightly rough texture on the branches. Even with a hand-lens, only the larger leaves on the main stems are visible, not those on the branches. The plants are green or yellow-green and dull rather than shiny. The main stem may be up to three inches long and has larger branches that are opposite each other. In turn each branch has smaller opposite branches. Fern Moss often grows in very large mats covering logs, soil, rocks, or even the bases of trees.

The commonest species of Thuidium, T. delicatulum, usually produces sporophytes in damp places in late fall. They are light reddish to golden brown in color and over an inch tall, with curved capsules. Each operculum has a moderately long "nose" on top. The peristome teeth are large enough to be easily noticeable. If you find Fern Moss sporophytes in winter when they are completely mature, bring a piece of the gametophyte with a few sporophytes indoors to watch the hygroscopic motion of the peristome teeth as they dry. The teeth spread apart in dry air so that spores can be released and then close again if returned to a more humid environment.

In shady moist areas near streams or seeps, Thuidium delicatulum is often the main bryophyte present. It is one of the commonest mosses in the state, and forms extensive mats, especially in areas of acid soil. Its new growth in spring and summer is quite green. In late fall and winter T. delicatulum is yellow-green.

The genus Thuidium includes several species besides T. delicatulum. The long drawn-out "nose" on its operculum and various details of its leaves distinguish this common species from T. recognitum, the most similar species of Fern Moss, which is found on calcareous rocks or on soil that is not acidic. Another genus with a somewhat similar pattern of fine branches and extremely small leaves is Hylocomium (see p. 57), which occurs less frequently in Pennsylvania. Thuidium is much flatter than Hylocomium, which may have several layers of main branches.

Conocephalum in winter, with Thuidium (see p.65) x2.5

Conocephalum in spring with new bright green growth x1

Conocephalum conicum

Though there is no common name for this large thallose liverwort, a good one might be the Snake-skin Liverwort, because of the scale-like pattern on its upper surface. It is the largest of the complex thallose liverworts that is common in Pennsylvania. Like other complex thallose liverworts it has air pores and air chambers, but it is the only one where the outlines of the chambers show on the surface. It grows along shady streams and on rocks in them, especially if they are spring-fed and do not dry up too often. It is about half an inch wide and may reach lengths of several inches. The thallus grows at the tip and from time to time the growing point divides in two so that the thallus forks into two equal parts. When you find Conocephalum, be sure to lift the tip of the thallus to see the fine hair-like rhizoids that fasten it down to its substrate. They show better in this species than in most other bryophytes.

When reproducing sexually, Conocephalum grows special branches with female sexual organs. These are cone-shaped and at first not raised much above the general level of the thallus. By the time the sporophytes are mature, however, the lower part of the branch elongates, often to an inch or more in height. This allows for more effective spore dispersal. The female branches and the sporophytes are not produced often, and the rounded low male reproductive structures are even less often seen. You may see only male or only female reproductive structures, since each thallus and the neighboring ones (which branched from the same original thallus) have either female or male reproductive structures, not both.

By late fall the Conocephalum thallus is a very dull green, and remains so in the winter. In spring when the growing tips become active again, the new growth is a bright grass green that contrasts very strikingly with the surviving last year's thallus.

Conocephalum grows on soil, rocks, or well rotted logs in wet areas, especially along or in streams and near waterfalls It is often growing on the tops of rocks where Porella (see p. 81) grows along the waterline. Along the margins of streams it may be found with Pallavicinia (see p.73) on moist soil.

Marchantia polymorpha showing gemmae cups x1.5

Marchantia – with young female reproductive branches x2

EMDE

With older female branches and the top of an old male branch x1

Marchantia polymorpha

Marchantia is a large thallose liverwort with a smooth upper surface. Its broad flat thallus grows at the tip and often divides into fairly equal branches. Even the smallest piece of the thallus is easily recognizable if it is reproducing asexually by means of round gemmae cups, as it often does. No other liverwort has round gemmae cups. (In the introduced genus Lunularia, which is sometimes found in greenhouses, the gemmae cups are crescent-shaped.) The gemmae (buds) in the cups are knocked out by raindrops and start new plants when they grow. For this reason the cups are also called splash-cups.

When the conditions are right for sexual reproduction in Marchantia, the thallus produces special upright branches where the reproductive organs are formed. On male plants these branches are short-lived and shaped like small mushrooms with flat tops, in which the sperm-producing organs are embedded. In female plants the reproductive branches look like tiny palm trees or, from the top, like an octopus. When these branches are young and not very tall the eggs are fertilized. Then as the sporophytes develop each branch elongates, often to an inch or more in height. The elevation of the sporophytes above the ground helps make the spore dispersal more efficient. When the sporophytes of Marchantia are fully mature, they look like small yellow eggs tucked under the top of the "palm tree." Thousands of yellow spores about to be released from the egg-shaped capsules provide the color.

Marchantia is found in nature on wet rock or soil, especially in areas that have recently burned. It is most commonly seen, though, in nurseries and greenhouses growing on rich soil around well-watered plants. Probably because of its tendency to grow in greenhouses, Marchantia is one of the best-known liverworts, certainly the one most often used as an example in botany textbooks. If you find a large thallose liverwort growing on wet rock, be sure to check whether the upper surface is smooth as in Marchantia or rough as in Conocephalum (see p. 67), which is much commoner in the forests of Pennsylvania.

If you find an even larger smooth liverwort on wet sandstone, it could be the elusive Dumortiera hirsuta, last seen in Pennsylvania in about 1905. This rare species has no gemmae cups, and its reproductive branches do not resemble palm trees.

Reboulia hemisphaerica x2

Reboulia – female branches with attached sporophytes, plus one male reproductive structure x1.5

Pellia epiphylla on a very steep shaded bank showing (at arrow) the site of female organs x2

P. epiphylla in winter with the purple color that it often shows after exposure to sun and cold x2.5

Reboulia hemisphaerica

This is a complex thallose liverwort with no gemmae cups. Its air-pores are not easily visible unmagnified. The thallus is a quarter inch or a little more in width and is lighter green than other liverworts, with dark purple edges. On the dark underside are rows of tiny crescent-shaped scales with two long teeth. Reboulia's reproductive structures are similar to those of Conocephalum (see p. 67), except that both types are on the same thallus. Its upright female branches have rounded tops with white hairs hanging down among the five egg-shaped sporophytes on each branch. (In the photo two or three are visible and the others are hidden.)

Reboulia grows on dry shaded rock outcrops, where it may form large patches. It grows best on calcareous rock, but also occurs on neutral or acid rock. At constantly wet calcareous sites a similar but slightly larger liverwort (with more obvious air-pores and tiny oval scales on the underside) is Preissia quadrata.

Pellia

A fairly large thallose liverwort without air-pores, growing in a moist place, is most commonly Pellia. The thallus has lobes that are about three-eighths of an inch wide. It is normally dark green or, if exposed to sun, sometimes a dark purple. Though thick and nearly opaque in the center, it is thin and translucent on the edges. Large numbers of long rhizoids extend down from the lower side of the thick central area. Be sure to distinguish Pellia from the more delicate simple thallose liverwort Pallavicinia (see p. 73) which has most of the thallus translucent.

Sexual reproduction takes place in early summer. Sperm are produced in small swellings along the center of the thallus. Egg-producing organs are formed under flaps near the tips of the lobes, and by late summer immature sporophytes may be found there.

Pellia grows in moist shady places, such as stream banks or seeps in the forest. On steep north-facing banks near streams it may form large patches. Of the two species of Pellia in Pennsylvania, the commoner is Pellia epiphylla, which is irregularly lobed. It has female and male reproductive organs on the same thallus. Pellia neesiana is more ribbon-like with narrow regular lobes and the two types of reproductive organs on separate plants.

Pallavicinia on the side of a rotten log x1.5

Pallavicinia upright reproductive structures x3

Carolyn Reider

Pallavicinia showing male reproductive organs x2.5

Pallavicinia lyellii

This liverwort is one of the smaller simpler thallose liverworts. It is easily recognized by its size and color and by the thickened strand or midrib that extends down the center of the thallus. Pallavicinia runs like a narrow dark bluish-green ribbon along damp soil near streams, growing at the tip and very rarely branching. The thallus is about a fifth of an inch wide and an inch or two long and attached to the substrate is by rhizoids growing down from the thickened central strand.

When conditions are right for sexual reproduction, eggs are produced by one Pallavicinia thallus and sperm by another. The sperm-producing organs are found in small groups covered by flaps on the upper layer of the thallus. The egg-producing organs and the immature sporophytes are protected by a green sleeve-like structure rising directly from the midrib. Several of these are shown in the center picture. At least from late fall till early spring these upright reproductive structures persist on the thallus, waiting for the warmth and moisture necessary for the sporophyte to make its rapid elongation before dispersing its spores. If you find a lot of Pallavicinia in late winter, take a little with immature sporophytes indoors and put it into a very humid atmosphere in a closed jar. The warmth will cause the sporophytes to grow taller so that you can see them. They have delicate, narrow, white stalks and dark capsules that are about three times longer than wide.

Pallavicinia grows on the banks of small shaded streams and on rotting logs across spring runs or seeps. The soil it grows on must be acidic and is usually high in humus. Where the soil and spring water are more alkaline it grows only on logs. Pallavicinia is often seen near Atrichum undulatum (see p. 21) on stream banks, or with Aulacomnium palustre (see p. 27) in acid seeps. Pallavicinia is very easily damaged by frost or drought. In the harsher seasons look for it in the most sheltered nooks tucked into the bank of a stream.

Another simple thallose liverwort often found growing in similar habitats, sometimes very close to Pallavicinia, is Pellia (see p. 71). Pallavicinia is narrower than Pellia and its distinct central thickened strand is much more noticeable than the gradual thickening in the center of the thallus of Pellia.

Bazzania trilobata – Note the typical branching pattern at the arrow. x2.5

Bazzania trilobata – Note the three teeth at the outer end of each leaf. x5

Bazzania trilobata

This large leafy liverwort has elliptical leaves with three large teeth at the end. The stems normally branch into two equal pieces in a Y-shaped pattern. Unlike most liverwort gametophytes, Bazzania trilobata usually grows nearly upright (to an inch or more in height), forming thick mats. Less often it grows flat, creeping along its substrate. Another peculiarity of B. trilobata is that the leaves bend down from the stems. On the side of the stem that leans a little towards the ground, branches of a different kind are produced that grow downward and appear to be roots. Actually, they have tiny leaves on them which means that they are stem branches. Where these downward growing branches touch the ground, they produce lots of rhizoids, which help hold the plant in place. Also, on the lower surface of the stem there is a third row of smaller leaves between the two main rows. The width of each branch of the stem with the two main rows of leaves on either side is between an eighth and a quarter of an inch.

Sporophytes of Bazzania trilobata are rarely produced, but if you have located a good patch of gametophytes, look for the sporophytes in late summer or early fall. They emerge from cylindrical reproductive structures on the lower side of the stem.

Bazzania trilobata grows in moist shaded ravines, often where small spring-fed streams provide a constant supply of water. It is more common farther north, and may form large mats, not just on soil, but on logs or rocks. It is commonest in forests where Hemlock, White Pine, and Yellow Birch are the most important trees.

Of the three species of Bazzania in Pennsylvania, B. trilobata is by far the largest and commonest. Its dark green color and completely leafy stems differentiate it from B. tricrenata, which is brownish, and from B. denudata, which has bare areas on the stems where leaves are missing. They are found on rock outcrops, usually at high elevations.

No other liverwort resembles Bazzania. Its large size, along with its three-toothed leaves that hang from the stem, make it stand out from all the others. It is also unusual in having the lower edge of each leaf covered by the next leaf in line down the stem, in contrast to Plagiochila asplenioides (see p. 83).

Carolyn Reider

Lophocolea heterophylla in winter x2

L. heterophylla in summer x6

Carolyn Reider

Lophocolea heterophylla with reproductive structures at right x3

Lophocolea heterophylla

This small leafy liverwort flourishes in the cooler moister seasons of the year on old logs in moist woods. The logs must be without bark and fairly well rotted and are often lying in in a seepy area or near a stream. Lophocolea heterophylla is pale yellow-green in the cooler seasons of the year and greener in summer. It often forms round colonies with growth near the edges and reproductive structures towards the center. The leafy growing stems creep and cling closely to the surface of the wood, while the reproductive structures stand upright. These structures are cylinders of material similar to the leaves and conceal egg-producing organs or, later, the immature sporophytes that result from sexual reproduction. The mature sporophytes with tall thin stalks elevating the capsules are very delicate, very short-lived, and rarely observed. In early spring the top of the dark-colored capsules are sometimes seen at the open upper end of the reproductive structures before the elongation of the delicate stalks occurs.

The general shape of the leaves of Lophocolea heterophylla is nearly rectangular. The lower edge of the leaf towards the tip of the stem overlaps the upper edge of the next leaf down the stem. With magnification it is sometimes possible to see that the leaves may be slightly bilobed at the end. The presence of both lobed and unlobed leaves on the same plant is the source of the species name heterophylla, meaning different leaves.

Lophocolea heterophylla is a very common liverwort, found not just on rotting logs in wet forests, but even in drier, disturbed forest or on abandoned woodpiles in shady places. It is most often seen on logs, but sometimes grows on smaller pieces of decaying wood or even on moist soil by a stream. Other species in this genus are less common, differ in leaf shape from L. heterophylla, and do not grow on wood.

One newer, more technical reference refers to this liverwort as Chiloscyphus profundus (the liverwort guide by Marie Hicks, see p. 88), but other experts still use the name given here.

<u>Nowellia</u> in early winter with white-tipped reproductive structures, center and left x4

<u>Nowellia</u> in summer, on the same log x4

Nowellia curvifolia

This is a very delicate leafy liverwort that coats old logs with a dark reddish color in the cooler seasons of the year. At 10x magnification it is possible to see that the leaves are divided into two narrow, sharply curved lobes, but the color alone on a log dead long enough to lose its bark is a sufficient identifying feature. Though leaves of most liverworts are flattened in one plane, Nowellia leaves are so curved that they stand out around the stems. They are very concave, like billowing sails in miniature. During the summer in deciduous forests, or in other very shaded locations, Nowellia is harder to recognize because its leaves are green. The hanging stems and the way their covering of curved leaves give them a cylindrical appearance suggest Nowellia, but for certain identification it is necessary to check that the leaves have two long narrow lobes.

When the weather is cool, reproductive structures may be seen on the central, older stems of the colonies of Nowellia, with low, creeping, non-reproductive stems growing out around them. The reproductive structures are cylindrical and open at the outer end, like a sleeve. They are mostly the same dark color as the rest of the plant, but may be lighter or even white at the tip, especially after they have been exposed to subfreezing temperatures. The egg-producing structures of the plant, and later the immature sporophytes, are located inside these tiny sleeves. Mature sporophytes appear in early spring and are very short-lived.

Nowellia is most often found on conifer logs, and always on dead logs that have lost their bark but are not completely rotten yet. It is commoner in the northern part of Pennsylvania than in the south. Sometimes Nowellia and Lophocolea heterophylla (see p. 77) are found on the same log.

The genus Nowellia was named after an early nineteenth century English hand-weaver who was an amateur bryologist and whose extensive collection of mosses and liverworts is still in the British Museum of Natural History.

Porella pinnata, dry, near a stream, at slightly less than life-size

Porella pinnata wet on a rock in a stream x1

Detail of the leaves of Porella pinnata
The arrow indicates the tip of a stem. x2.5

Porella pinnata

This large leafy liverwort grows on rocks. At first glance, leaves of Porella pinnata seem to be round or elliptical and unlobed. Its stems are fastened to the substrate near the base, but usually hang free along most of their length (as in the center picture on the facing page). The plants are dark green and translucent when wet, but an opaque dark green when dry. Each leaf is attached on a slant so that its upper edge overlaps the lower edge of the leaf next above it on the stem. With good light and an 8x hand-lens to examine the underside of the stems, it is clear that the leaves actually have two lobes, with the large rounded lobe obvious from above or below and the small, narrow, tongue-shaped lobe visible only from below. It also may be possible to see that there is a row of smaller nearly round leaves along the underside of the stem between the two rows of ordinary leaves. The width of the stem with the two rows of leaves is more than one-eighth inch. The plants are up to four inches long.

In Porella pinnata the protective structure around the female reproductive organs is like a rounded sac, somewhat longer than wide. After fertilization has taken place, the immature sporophytes remain hidden inside till spring, when the stalks elongate, the capsules split, and the spores are released. In contrast to most other liverworts, the sporophytes of P. pinnata are substantial enough to remain attached to the gametophytes after the spores have been distributed. The old, dry sporophytes are visible for months afterwards, topped by the four narrow dark brown sections of the empty capsule. The sac-like female branch and the sporophyte emerging from it are at least an eighth of an inch long

Porella pinnata grows on rocks that are wet, either constantly or intermittently. It is often found in spring-fed streams forming a ring around each rock, either at the water-line or submerged just below the water's surface. Less commonly, P. pinnata grows on rock faces (usually vertical ones) on damp shaded areas near streams. In streams it sometimes grows along the sides of the same rocks where Conocephalum (see p. 67) is on the upper surface. A large liverwort growing on rock with leaves nearly round in appearance is almost always P. pinnata. Finding the tongue-shaped smaller lobes of the leaves then confirms the identification of this species.

Plagiochila asplenioides on a rock in a stream x3
The leaf bases extend down the stem (see arrow).

Scapania nemorosa showing the leaf arrangement x4

S. nemorosa with gemmae x5.5

Plagiochila asplenioides

This is one of the largest leafy liverworts. Each stem of the gametophyte with its two rows of leaves is an eighth of an inch or more wide and usually half an inch or more long. The stems are attached only at the base and grow nearly upright or sometimes hang down. Each leaf is elliptical in shape, except that the corner nearest the base of the stem extends down along it. With an 8x lens, fine teeth are often visible on the edge of the leaf. The attachment of the leaf is slanted on the stem so that the corner extending down along the stem overlaps the top of the leaf next in line. Since each stem resembles a small simple fern such as Ebony Spleenwort, and liverworts are sometimes referred to as hepatics, A. J. Grout in his *Mosses with a Hand-lens* (see p. 88) gave this species the common name Spleenwort Hepatic.

The habitat for Plagiochila asplenioides is moist to wet. It often grows in clumps, its nearly upright stems with their bases in damp sand or silt, near a stream. It may grow on damp rocks or hang from a sandy rock in a stream as in the photo on the facing page. (See also the upper photo, p. 6.) The combination of its large size, the shape and attachment of its leaves, and its wet or seasonally wet and shady habitat make this species easy to recognize. Three other species of Plagiochila grow in Pennsylvania but P. asplenioides is the largest and most distinctive.

Scapania

In the genus Scapania the leaves have two lobes, with the smaller lobe folded over the larger. Thus, the plants appear to have four rows of leaves instead of two, with two rows of larger leaves extending on either side of the stem and two rows of smaller leaves in the center covering the stem. The leaves are green or green tinged with red. Groups of tiny dark gemmae are often found at the stem tips. Though there are several species of Scapania in the state, the largest and commonest is S. nemorosa, which has fine teeth on the edges of its leaves.

Scapania is usually found in wet habitats on sand, soil, or rock near streams or seeps. In the other genus of liverworts with the smaller leaf lobes on top (Diplophyllum), the plants are tiny, and the leaves are longer and narrower than in S. nemorosa.

Ptilidium pulcherrimum
with bronze coloration x5

P. pulcherrimum with
immature capsules x7

Carolyn Reider

Trichocolea tomentella x3

T. tomentella with Thuidium delicatulum x4

Ptilidium

In the genus Ptilidium, the delicate leaves are divided into two main sections that are fringed with narrow hair-like lobes. The plants branch extensively and may form large mats, which are pale green to greenish bronze or bronze tinged with dark red. New growth is greener, becoming a yellower green and finally in cold weather developing a reddish tinge if light levels are moderately high. Because the plants are so small, less than a tenth of an inch wide with the two rows of leaves, the color of the mat is often a clue to its identity. The location of the liverwort helps too. Ptilidium is often found in drier habitats than most other liverworts, or at least in habitats that are seasonally dry. It does not require a spot with constant moisture, as the larger (also hairy) liverwort Trichocolea does.

Of the two species of Ptilidium found in Pennsylvania, P. ciliare grows only in northern conifer forests. P. pulcherrimum is more common and grows in diverse forests on moist decaying logs or less often on acid soil or rocks. Its mats are very thin and attached closely to the substrate, while those of P. ciliare are thicker, with the stems growing more upright. When sexual reproduction occurs in P. pulcherrimum, the resulting sporophytes mature in early spring. After releasing the spores, the dry sporophytes with empty capsules may be visible for several months.

Trichocolea tomentella – Woolly Liverwort

This large liverwort has leaves divided and redivided into segments so fine as to be hair-like. The stems branch rather regularly in a fern-like pattern. A main stem with all its side branches is often an inch or more across. Though sporophytes are rarely produced, keep an eye out for them in spring.

Trichocolea forms thick, light green patches in very moist shaded locations on the edges of seeps or streams. It does not grow in places that ever dry out. Often it grows with the moss Thuidium (see lowest photo, left, and p. 65), which has a similar pattern of branching but is finer and greener. In winter the fuzzy leaves of Trichocolea are almost white, which accounts for its common name. Both the genus name and the species name of this liverwort have roots referring to hair. Tricho- is from the Greek for hair, and toment- is from the Latin.

GLOSSARY

Also see labeled illustrations of moss and liverwort terminology inside the front and back covers.

Air chambers – Spaces in the upper part of the thallus of complex thallose liverworts where photosynthesis takes place.

Air pores – Openings allowing oxygen and carbon dioxide to move between the air chambers and the air above the thallus.

Asexual reproduction – Reproduction of the gametophyte of a moss or liverwort directly, without the sexual process and without sporophytes or spore production.

Bryophyte – A collective term for mosses and liverworts. Both are small land plants with relatively large gametophytes normally not having the ability to take up water from the soil. Their sporophytes are attached to their gametophytes.

Calcareous – Calcium-rich and non-acidic.

Calyptra – A protective covering over the developing sporophyte. It is derived from the egg-producing organ of the gametophyte.

Capsule – The upper, swollen, spore-producing portion of the sporophyte of both mosses and liverworts.

Cuticle – A layer of water-proof wax secreted on the surface of higher plants by the outermost layer of cells.

Egg – The female reproductive cell, which must be fertilized by a sperm before growing into the sporophyte.

Fertilization – The fusion of egg and sperm.

Gametophyte – The green stage of a bryophyte that can continue to grow and to produce gametes (eggs and sperm).

Gemmae – Small bud-like structures used in asexual reproduction of gametophytes of mosses and liverworts.

Lobed – Having projections or divisions. A lobed leaf is at least shallowly divided into two or more sections.

Midrib – A thickened central strand which supports a moss leaf and resembles a vein, but does not carry water.

Mineral soil – Soil with little organic material or humus.

Operculum – The cover that falls off the top of a moss capsule, allowing the spores to sift out between the peristome teeth. It may be flat or conical and often has a tall projection from its center, like a long nose or handle.

Protonemata – Filaments of cells that grow from moss spores. They are a very young form of the moss gametophytes.

Peristome teeth – Structures found near the top of a moss capsule but under the operculum. They regulate spore release, in most cases by spreading apart when the air is dry.

Reproductive branches – Upright portions of the thallus of complex thallose liverworts where egg- or sperm-producing organs are located.

Rhizoid – A very thin filament of cells which fastens the lower part of a moss or liverwort gametophyte to its substrate.

Sperm – A male reproductive cell that fuses with an egg (i. e. fertilizes it) to produce a zygote.

Splash cup – A dish-shaped structure the right size to capture the force of a rain drop to disperse gemmae or sperm.

Spore – A cell that can grow directly into a new plant.

Sporophyte – The upper, short-lived part of a bryophyte which produces spores and is normally not green when mature.

Stalk – The narrow lower part of a moss or liverwort sporophyte, which normally is tall and elevates the capsule.

Substrate – The material the moss or liverwort grows on, usually soil, wood, or rock.

Thallus – A plant structure, as in some liverworts, that is not differentiated into stems and leaves.

Thallose – An adjective referring to a thallus.

Zygote – A cell resulting from the fusion of egg and sperm, i.e., a fertilized egg.

BIBLIOGRAPHY

Henry S. Conard and Paul S. Redfearn, Jr., 1979, *How to Know the Mosses and Liverworts*, Wm. C. Brown, Dubuque, Iowa

Howard A. Crum and Lewis E. Anderson, 1981, *Mosses of Eastern North America*, Columbia University Press, New York, New York

Howard A. Crum, 1973, *Mosses of the Great Lakes Forest*, University of Michigan, Ann Arbor, Michigan

Janice M. Glime, 1993, *The Elfin World of Mosses and Liverworts of Michigan's Upper Peninsula and Isle Royale*, Isle Royale Natural History Association, Houghton, Michigan

A. J. Grout, 1924, *Mosses with a Hand-lens*, 3rd ed., New York, New York, published by the author

Marie L. Hicks, 1992, *Guide to the Liverworts of North Carolina*, Duke University Press, Durham, North Carolina

C. B. McQueen, 1990, *Field Guide to the Peat Mosses of Boreal North America*, University Press of New England, Hanover, New Hampshire

G. Christine Manville, 2001, *An Annotated Checklist of Liverworts and Hornworts of Pennsylvania*, Philadelphia, published by the author with the support of the PA DCNR

Rudolf M. Schuster, 1949, The Ecology and Distribution of Hepaticae in Central and Western New York, *American Midland Naturalist*, Vol. 42 (No. 3): pp. 513-712

Rudolf M. Schuster, 1966-1992, *The Hepaticae and Anthocerotae of North America*, vols. 1-6, Columbia University Press, New York, New York

INDEX OF NAMES OF MOSSES AND LIVERWORTS

MOSS TERMINOLOGY

Moss Leaves

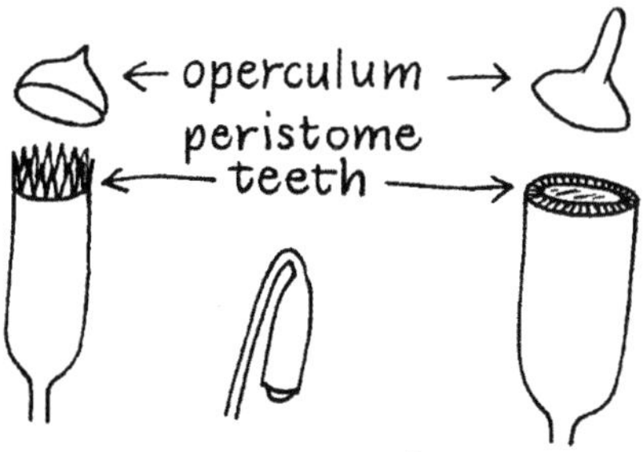

Moss Capsules

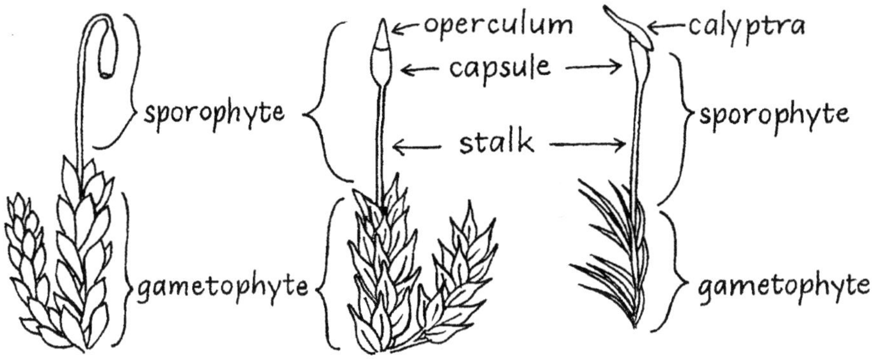

Acrocarpous Mosses

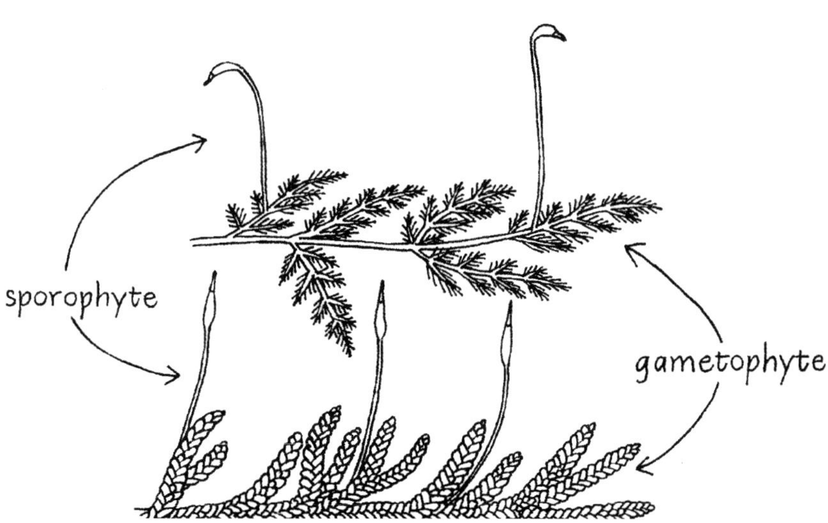

Pleurocarpous Mosses